Process Intensification

Topical Conference at the 2022 AIChE Spring Meeting and
18th Global Congress on Process Safety

San Antonio, Texas, USA
10 – 14 April 2022

ISBN: 978-1-7138-6824-8

Printed from e-media with permission by:

Curran Associates, Inc.
57 Morehouse Lane
Red Hook, NY 12571

Some format issues inherent in the e-media version may also appear in this print version.

Printed with permission by Curran Associates, Inc. (2023)

For permission requests, please contact AIChE
at the address below.

AIChE
120 Wall Street, FL 23
New York, NY 10005-4020

Phone: (800) 242-4363
Fax: (203) 775-5177

www.aiche.org

Additional copies of this publication are available from:

Curran Associates, Inc.
57 Morehouse Lane
Red Hook, NY 12571 USA
Phone: 845-758-0400
Fax: 845-758-2633
Email: curran@proceedings.com
Web: www.proceedings.com

TABLE OF CONTENTS

Redefining the Upper Bound Curve for Process Intensification in Natural Gas Processing

Paper #: 22a

Jimoh K. Adewole[1], Naveed Khan, Asma S. Al Kharusi, Muhammad Javed, Habeebllah B. Oladipo, Muhammad Zahir Khan, Riaz Ahmad

Membrane Science and Engineering Lab, Department of Engineering (Process Engineering), International Maritime College, Sultanate of Oman

Extended Abstract

The use of membrane for acid gas (such as CO_2) separation has attracted significant attention since its ranking as the twelfth out of the 38 research topics constituting the research needs in the membrane separation industry. Though, membrane application has witnessed about tenfold growth afterwards, however, solvent-based absorption still has about 90% of the natural gas (NG) processing market. This is because today's membranes are not good enough to displace the use of absorption. High pressure operations have been suggested as one of the means to improve the competitiveness of membrane for separation of acid gases. Unfortunately, the penetrant-induced plasticization phenomenon has limited the use of membrane for high-pressure CO_2 separation. A number of studies on plasticization resistant membrane material and the optimization of membrane-based gas separation processes have been published. Most of these studies have employed models with constant CO_2 permeability coefficients while evaluating high-pressure gas separation.

The present paper proposes the use of a process intensification approach that takes into account the variation of gas permeability with upstream pressure. Therefore, an approximate model based on plasticization pressure and permeability parameters at plasticization was proposed for defining the Robeson upper bound curve of membranes used for high pressure gas separation. The model which was derived by applying the partial immobilization assumption to the fundamental model of solution - diffusion was used for selecting the best membrane material which was used in this work. A comparative study of the traditional solvent-based CO_2 separation and membrane-based gas separation techniques was carried out on a gas separation plant using the process intensification approach. The study was performed using process intensification metrics including mass intensity; waste intensity; productivity/size ratio; productivity/weight ratio; flexibility and modularity and a multi-objective optimization method. The results obtained indicated that huge saving is possible by installing a plasticization resistant membrane in an area which involves high pressure CO_2 separation. Furthermore, this idea is justified since natural gas is often obtained directly from gas wells at pressure ranging from 20 to 83 bar and compositions between 4 and 50% CO_2.

---

[1]jimoh@imco.edu.om (Jimoh K. Adewole)

643277 Process Intensification and Reactor Design for Efficient SAF Production

Abstract:

Bioenergy continues to attract considerable attention as a potential solution to national and global concerns regarding climate change, sustainable energy, and energy independence/security. Indeed, the US Department of Energy has recently released a Sustainable Aviation Fuel (SAF) Grand Challenge with aggressive targets in 2030 and 2050 for the domestic production of SAF. Meeting this challenge will require the development of new materials and processes along with the underlying process science and engineering to enable to scale up and deployment of these technologies.

In this paper, we will review in situ catalytic fast pyrolysis (CFP) as an effective technology for the production of SAF and other fuels and chemicals. The key considerations between in situ and ex situ CFP will be discussed from a process intensification perspective. Furthermore, modifications of NREL's thermochemical pilot plant to enable piloting of in situ CFP technologies, along with some key challenges and developing solutions will be presented.

639043 Process Intensification of Enzyme Catalysed Transesterification of Geraniol in the Spinning Mesh Disc Reactor

Abstract:

In recent years, chemical industries have begun to move towards more sustainable processes and practices to meet environmental regulations. Global political and pandemic events have exposed the inability of industrial processes to respond effectively to changes in demand, highlighting the importance of localised and resilient manufacturing. Reactor technologies that have shown potential for achieving process intensification through flexible, resource and energy efficient manufacturing are fast gaining commercial attention over traditional batch reactors. The spinning mesh disc reactor (SMDR) is one such process intensification technology that uses centrifugal force to drive reaction fluid over a mesh-supported catalyst on a rotating disc [1, 2]. The SMDR is modular, flexible and scalable and has shown be 50% more resource efficient compared to traditional batch reactors through improved productivity and re-usability of expensive catalysts like enzymes [3, 4]. In this study, we have demonstrated the potential of the SMDR for intensification of enzymatic structural modification of terpene alcohols into terpene esters.

The lipase-catalysed production of geranyl acetate was first investigated in batch, using both free lipase and lipase immobilised on woollen cloth. The synthesis pathway chosen was transesterification using vinyl acetate in ethyl acetate solvent. Using a design of experiments approach, significant variables affecting the reaction were investigated. Of the variables studied, temperature and molar ratio were determined to be the most impactful, with catalyst content and use of immobilisation being much less significant. Further batch experiments were also performed to investigate other parameters such as stirring speed, acyl donor concentration and substrate concentration. Stirring speed was observed to be significant for free lipase from 225 to 400 RPM, but insignificant for those same ranges for immobilised lipase. Acyl donor concentration was seen to be directly proportional with reaction rate for the range studied (0.1 to 0.4M of vinyl acetate with 0.1M of geraniol). The effects of increasing substrate concentrations were also investigated, with both constant and increasing catalyst concentrations. For free lipase, reaction rates were observed to increase from 0.01 Mh^{-1} to 0.07Mh^{-1} with increasing catalyst concentration (w/w of initial substrate concentration), demonstrating a synergistic relationship between catalyst content and substrate concentration. This synergy was not seen however for immobilised lipase.

The optimum reaction conditions were identified using design of experiments in batch and a maximum reaction conversion of 80% was observed at 40°C at the end of 24 hours. The reaction was then scaled-up in the SMDR and a reaction conversion of 60% was obtained at 350 RPM and 3 ml s^{-1} in 5 hours. Further, the catalyst loading in the reactor was increased by simply adding more cloths on the surface of the disc and a 5% increase in the reaction conversion was observed with the addition of two cloths. The lipase cloth was re-used for up to three cycles and 83% of the original activity was retained, demonstrating the robustness of this catalyst for organic reactions.

To the best of the authors' knowledge, this is the first example of scale-up of production of terpene esters in rotating disc reactors. We are currently working towards further optimising the SMDR to improve the reaction productivity for pilot scale operations.

References

[1] X. Feng, D. A. Patterson, M. Balaban, G. Fauconnier and E. A. C. Emanuelsson, *Chem. Eng. J.*, 2013, **221**, 407-417.

[2] P. Shivaprasad, M. D. Jones, D. A. Patterson and E. A. C. Emanuelsson, *Chemical Engineering and Processing: Process Intensification*, 2017, **122**, 550-559.

[3] E. A. C. Emanuelsson, A. Charles and P. Shivaprasad, *Sustainability*, 2021, **13**, 6944.

[4] P. Shivaprasad, M. D. Jones, D. A. Patterson and E. A. C. Emanuelsson, *Chemical Engineering and Processing-Process Intensification*, 2018, **132**, 56-64.

638960 Kinetic Resolution of 1-Phenylethanol in the Spinning Mesh Disc Reactor: A Techno-Economic Analysis

Abstract:

An increasingly important reaction in the pharmaceutical and fine chemicals industry is the kinetic resolution of racemates; a result of the demand for chiral intermediates growing from both pharmaceutical and agricultural industries [1]. An example of this is the kinetic resolution of 1-phenylethanol. 1-phenylethanol can undergo an acylation reaction when in the presence of an enzyme that produces the ester, R-1-phenylethyl acetate at a rate much greater than the optically opposite ester. This allows for the separation of the two enantiomers to be used in their respective applications.

Process intensification is generally a development that leads to a more environmentally and economically efficient process [2]. For the kinetic resolution reaction, this has come in a multitude of forms. One solution is the spinning mesh disc reactor (SMDR). Developed from the concept of a spinning disc reactor (SDR), with the addition of a mesh housing purposed for the immobilisation of enzymes, the SMDR has shown to achieve process intensification [3]. The advantages of the SMDR in comparison to a traditional batch reactor lie in its enhanced mass transfer (due to the thinly sheared film of reactant fluid spread across the rotating discs surface), residence time distribution and ability to protect the enzymes from shear forces [4]. It also has faster mixing and can reduce the need for large quantities of enzyme by increasing reusability in comparison to a batch reactor using free enzyme. So far, the SMDR has shown success on the laboratory scale but no research has been conducted on the economic feasibility of the reactor. Techno-economic analysis (TEA) is a tool that can determine economic viability beyond the lab scale and sensitivity analysis can be used to determine the uncertainty in the model. Within this study, we have carried out a comparative TEA of the kinetic resolution of 1-phenylethanol in a batch reactor and the SMDR.

Two production scales of 100 and 20 tonnes per year were considered – the former being the most competitive economically. Energy consumption in the SMDR was higher than that of the batch process, requiring 120,000 kWh per year compared to 19,000 kWh at the 100 t/y scale. The difference was almost five times greater at the smaller scale. Uncertainty in the process design and prices used for modelling cost was high. Therefore, probabilistic estimates of the cost of manufacture (COM) were generated using Monte Carlo simulations. At both scales, variation in COM was higher in the batch reactor. For 100 t/y, a 50% confidence interval surrounding the mean confirmed the COM for the SMDR would be lower than that of the batch reactor. The batch process had a 50% probability of costing between £18,433 and £21,229 per tonne of product. For the SMDR the range fell between £16,413 and £18,171. At 20 t/y, the batch and SMDR processes had significant overlap of estimated costs and a much lower probability that the SMDR would be cheaper than the batch reactor. Minimum estimated selling price (MESP) was determined to be most sensitive to the conversion of the reaction and the cost of raw materials. Specifically, the batch process was most sensitive to lipase cost (as 11 times the amount was required compared to the SMDR) and the SMDR to 1-phenylethanol cost.

To the best of the authors' knowledge, this is the first study on the economic feasibility of using the novel SMDR for kinetic resolution of 1-phenylethanol. The SMDR has clear economic advantages derived from its enhanced mass transfer and enzyme immobilisation, in comparison to traditional batch reactor at large scales. The implementation of the SMDR in industry for reactions of this nature is likely to be beneficial in both long and short-term applications.

References:

[1] M. Habulin and Z. Knez, "Optimization of (R,S)-1-phenylethanol kinetic resolution over Candida antarctica lipase B in ionic liquids," *Journal of Molecular Catalysis B: Enzymatic,* vol. 58, pp. 24-28, 2009.

[2] G. Keller and P. Bryan, "Process engineering: moving in new directions," Chemical Engineering Progress, pp. 41-50, 2000.

[3] P. Shivaprasad, M. D. Jones, D. A. Patterson and E. A. C. Emanuelsson, "Kinetic resolution of 1-phenylethanol in the spinning mesh disc reactor: Investigating the reactor performance using immobilised lipase catalyst," Chemical Engineering and Processing - Process Intensification, vol. 132, pp. 56-64, 2018.

[4] P. Oxley, C. Brechtelsbauer, F. Ricard, N. Lewis and C. Ramshaw, "Evaluation of Spinning Disk Reactor Technology for the Manufacture of Pharmaceuticals," Industrial & Engineering Chemistry Research, vol. 39, no. 7, pp. 2175-2182, 2000.

640152 Comparison of the Dual-Process Langmuir and Generalized Langmuir Isotherms for Representation of Unary, Binary, and Ternary Equilibria

Abstract:

Mixed gas adsorption is still a very active area of research, where newly proposed models need to be compared actively to standards in the field. The generalized Langmuir (gL) isotherm, proposed by Hamid *et al.* to address adsorption of different size molecules on heterogeneous adsorbents without requiring evaluation of the Gibbs isotherm, [1] has thus-far been tested mainly against the adsorbed solution theory (AST). The gL isotherm was shown to be thermodynamically consistent, [2] and gave either better or similar representation than both IAST and RAST-aNRTL. [1 - 3] Here, the gL isotherm will be contrasted with the duel-process Langmuir (DPL) isotherm popularized by Ritter and his coworkers at USC. [4 - 7]. The models will be compared based on their representation of the equilibrium loading and isosteric heat of adsorption (where available) of individual components, binary, and ternary mixtures of various gaseous adsorbates on different adsorbents over a range of experimental pressures and temperatures.

Works Cited:

1. Hamid, U.; Vyawahare, P.; Tun, H.; Chen, C.-C. (2021). Generalization of Thermodynamic Langmuir Isotherm for Mixed-Gas Adsorption Equilibria. *In Review.*
2. Hamid, U.; Vyawahare, P.; Sees, M.D.; Chen, C.-C. (2021, November). "Generalized Langmuir Isotherm Addresses Spreading Pressure Dependency of Modified Ideal Adsorbed Solution Theory." Paper presented at the 2021 Annual Meeting of the American Institute of Chemical Engineers, Boston, MA.
3. Hamid, U.; Vyawahare, P.; Chen, C.-C. (2021, November). "Isosteric Heat of Adsorption of Pure Component and Mixed-Gas Adsorption from Generalized Langmuir Isotherm." Paper presented virtually at the 2021 Annual Meeting of the American Institute of Chemical Engineers, Boston, MA.
4. Ritter, J.A.; Bhadra, S.J.; and Ebner, A.D. (2011). On the Use of the Duel-Process Langmuir Model for Correlating Unary Equilibria and Predicting Mixed-Gas Adsorption Equilibria. *Langmuir*, 27, 4700 – 4712.
5. Bhadra, S.J.; Ebner, A.D.; Ritter, J.A. (2012). On the Use of the Duel-Process Langmuir Model for Predicting Unary and Binary Isosteric Heats of Adsorption. *Langmuir*, 28, 6935 – 6941.
6. Ritter, J.A.; Bumiller, K.C.; Tynan, K.J.; and Ebner, A.D. (2019). On the Use of the Duel-Process Langmuir Model for Binary Gas Mixture Components that Exhibit Single-Process or Linear Isotherms. *Adsorption*, 25, 1511 – 1523.
7. Tynan, K.J.; Tosso, S.; Ebner, A.D.; and Ritter, J.A. (2021). On the Use of Single, Dual, and Three-Process Langmuir Models for Binary Gas Mixtures that Exhibit Unique Combinations of These Processes. *Adsorption*, 27, 637 – 658.

640072 Modeling Aqueous Solutions of Rare Earth Elements (REE) Salts with Association Electrolyte Nonrandom Two-Liquid Activity Coefficient Model

Abstract:

This paper presents an extension of the association electrolyte nonrandom two-liquid (AeNRTL) activity coefficient model [1] to calculate the activity coefficients for aqueous salt solutions of rare-earth elements (REE). Given the extensive consumption and limited supply of REEs in electronics, clean energy, aerospace, automotive and defense, the separation and recovery of REEs are of national importance [2]. Rigorous thermodynamic modeling for aqueous salt solutions of REEs is a key enabling technology in the research, development, design, and simulation of chemical processes for the separation and recovery of REEs.

Development of comprehensive thermodynamic models for the aqueous salt solutions of REEs has been very limited in the literature due to the lack of experimental thermodynamic data and the lack of rigorous electrolyte thermodynamic models capable of modeling ions of exceedingly high surface charge density including the REE ions. The recent advent of the AeNRTL activity coefficient model offers a major breakthrough in electrolyte thermodynamics especially for aqueous electrolyte solutions with high surface charge density ions [1]. This work reports compilation of the appropriate experimental thermodynamic data and accurate representation of these thermodynamic data for aqueous salt solutions of REEs and their mixtures with AeNRTL. The resulting model should be a valuable tool in support of the development of efficient and robust separation and recovery processes for REEs.

640149 Modeling Adsorption of Organics-Laden Air from Wood Drying Using Equilibrium Approaches

Abstract:

Air leaving a wood dryer contains organics which must be removed before the air can be released to the atmosphere. The current state-of-the-art in wood drying is the regenerative thermal oxidizer (RTO), which reduces the organics to carbon dioxide over a packed bed of ceramics (and sometimes iron catalyst) at high temperatures. [1,2] Sustainability and economic pressures suggest a more passive mode of capture, where the organics are condensed and recovered instead of burned. Captis Aire has developed a moving bed thermal adsorption process which can recover organics from air using a fluidized bed concentrator (FBC). A process model was developed in AVEVA Simulation which models the FBC as a series of equilibrium adsorption stages, as well as modeling the downstream desorption and condensing operations. While a traditional analysis using Murphree tray efficiencies to account for discrepancies is certainly possible, the recently proposed virtual moving bed model (VMB) suggests a different efficiency based on the equilibrium loading is also available. [3] The behavior of both types of efficiencies are reviewed. Additionally, the adsorption of α-pinene, a representative compound for a collection of terpenes, was modeled with two different isotherm forms: The Langmuir-Freundlich (Sips) isotherm and the thermodynamic Langmuir (tL) isotherm. [4–7] The capability of each model to represent the data is contrasted and recommendations are made for their use in commercial process simulators.

Works Cited:

1. Milota, M. R., and Western Dry Kiln Association. (2000). Regulation and Control of Air Emissions.
2. Zhang, Z., Jiang, Z., and Shangguan, W. (2016). Low-temperature Catalysis for VOCs Removal in Technology and Application: a State-of-the-Art Review. *Catalysis Today*, *264*, 270 – 278.
3. Sees, M.D., Kirkes, T., and Chen, C.-C. (2020). A Simple and Practical Process Modeling Methodology for Pressure Swing Adsorption. *Computers & Chemical. Engineering,* 2021, 147, 107235.
4. Chang, C.K., Tun, H., and Chen, C.-C. (2020). An Activity-Based Formulation for Langmuir Adsorption Isotherm. *Adsorption*, 2020, 26(3), 375 – 386.
5. Yang, R. T. (1997). *Gas Separation by Adsorption Processes*(Vol. 1). World Scientific.
6. Kaur, H., Tun, H., Sees, M.D., and Chen, C.-C. (2019). Local Composition Activity Coefficient Model for Mixed-Gas Adsorption Equilibria. *Adsorption*, *25*(5), 951-964.
7. Tun, H., and Chen, C.-C. (2020). Prediction of Mixed-Gas Adsorption Equilibria from Pure Component Adsorption Isotherms. *AIChE Journal*, 1 – 9.

639069 Model-Based Design and Intensification of a Catalytic Fluidized Bed Membrane Reactor for Oxidative Coupling of Methane

Abstract:

The catalytic oxidative coupling of methane (OCM) process has received growing scientific and commercial interests during the past decades. It offers the potential to directly convert methane to higher hydrocarbons (e.g., ethylene) from natural gas, and consequently reducing the cost, energy consumption, and carbon emissions. However, conventional industrial reactors (e.g., packed bed reactor and fluidized bed reactor) for this process suffer from low ethylene yields, high operating temperatures, and fast catalyst degradation, which hinder the commercialization of this technology (Tiemersma et al., 2012). To drive the innovation in OCM processes, a model-based reactor design approach can be useful for generating the optimal reactor design accounting for novel reactor concepts and benchmark with conventional process solutions.

The aim of this work is to present an optimal OCM process at commercial scale leveraging the concept of modular process intensification. An intensified fluidized bed membrane reactor (FBMR) catalyzed by La_2O_3/CaO has been investigated, and its performance is compared to a conventional fluidized bed reactor (FBR). Although the utilization of membrane for oxygen feed distribution adds to the capital cost investment and scaling up challenges, it has been reported to result in better ethylene yield and selectivity by selectively enhancing the desired reactions. To systematically identify the optimal design solution, high fidelity FBMR and FBR models are developed in gPROMS ModelBuilder, incorporating the partial differential algebraic equations accounting for mass balances, hydrodynamics, catalyst solid distribution, etc. A micro-kinetic model (with more than 350 elementary reactions) (Dooley et al., 2011) and a 10-step reduced kinetic model (Cruellas et al., 2020) are compared to study the impact of kinetics on reactor modeling and design . The two reactor configurations are optimized to obtain maximum profit, taking into account the design parameters on reactor sizing, temperature, catalyst particle velocity, and membrane tube design (for FBMR). Extensions of the reactor models with operational optimization following the PARametric Optimization and Control (PAROC) framework will also be discussed (Pistikopoulos et al., 2015).

References

Cruellas, A., Melchiori, T., Gallucci, F., & van Sint Annaland, M. (2020). Oxidative Coupling of Methane: A Comparison of Different Reactor Configurations. *Energy Technology*, *8*(8), 1–15. https://doi.org/10.1002/ente.201900148

Dooley, S., Burke, M. P., Chaos, M., Stein, Y., Dryer, F. L., ZHUKOV, V. P., FINCH, O., SIMMIE, J. M., & CURRAN, H. J. (2011). Methyl Formate Oxidation: Speciation Data, Laminar Burning Velocities, Ignition Delay Times, and a Validated Chemical Kinetic Model. *International Journal of Chemical Kinetics*, *43*(3), 154–160. https://doi.org/10.1002/kin

Pistikopoulos, E. N., Diangelakis, N. A., Oberdieck, R., Papathanasiou, M. M., Nascu, I., & Sun, M. (2015). PAROC—An integrated framework and software platform for the optimisation and advanced model-based control of process systems. *Chemical Engineering Science*, *136*, 115–138. https://doi.org/10.1016/J.CES.2015.02.030

Tiemersma, T. P., Chaudhari, A. S., Gallucci, F., Kuipers, J. A. M., & van Sint Annaland, M. (2012). Integrated autothermal oxidative coupling and steam reforming of methane. Part 2: Development of a packed bed membrane reactor with a dual function catalyst. *Chemical Engineering Science*, *82*, 232–245. https://doi.org/10.1016/j.ces.2012.07.047

640040 Modeling Aqueous Lithium Salt Systems with Association Electrolyte Nonrandom Two-Liquid Activity Coefficient Model

Abstract:

The advancement of process design, development, and simulation of electrolyte systems fundamentally lies within rigorous thermodynamic modeling. With high surface charge density, lithium ion exhibits strong hydration strength and results in exceedingly high solution non-ideality in aqueous lithium salt solutions. In this work, the recently developed association electrolyte nonrandom two-liquid (AeNRTL) model[1] is used to establish a comprehensive thermodynamic model for aqueous solutions of LiCl, LiBr, LiI, $LiNO_3$, and their mixed-salt systems. The solution non-ideality of the system due to ion hydration and ion-pairing is accounted for with the ion-specific association strength parameters while the solution non-ideality due to van der Waals physical interactions is captured with the binary water-salt and salt-salt interaction parameters. With the association strength parameters and the binary interaction parameters regressed from experimental data, we show the AeNRTL model provides the first-ever accurate engineering correlation for the available thermodynamic property and solubility data of aqueous lithium salt solutions over salt concentrations up to saturation and temperatures up to 523.15 K.

[1] Y. Lin, C.-J. Hsieh, C.-C. Chen, "*Association-based Activity Coefficient Model for Electrolyte Solutions*," *AIChE Journal*, 2021, e17422. Article DOI: 10.1002/aic.17422

AUTHOR INDEX